My Life As A Beekeeper

My Life As A Beekeeper

John A. George

A Bookland Juvenile

Carlton Press, Inc. *New York, N.Y.*

Manufactured in the United States of America
ISBN-0-8062-0963-1

My Life As A Beekeeper

This is a story about a little Hungarian boy, who comes from a poor family and wants to become a beekeeper. His father cannot help John in any way. This took place in 1936 when times were bad throughout the country and money was scarce. John's father told him to work the next two summers and in 1938, if he has enough money, could buy his bees and equipment. The first summer, John went from one farm to another earning twenty-five cents an hour, whenever the farmers had any work for him.

John was a little boy who loved to watch the bees on the flowers.

Each summer when John had a chance to he went to the farms to work, so that he could save what he earned for the Bee Hive.

Whatever money John could save, he would put in a glass jar, then hide away in the attic. This way he would not spend it on candy. After two summers of saving money, he had enough to buy his bees. When spring came in April of 1938, he told his father, "I am ready, Dad, to send a check off to the company who handles bees and equipment." His father was amazed, thinking that after two years his son forgot about beekeeping. John's father turned to his wife, who in turn said, "You told the boy he could have bees if he saved his own money." Now, John had the OK of his mother and father.

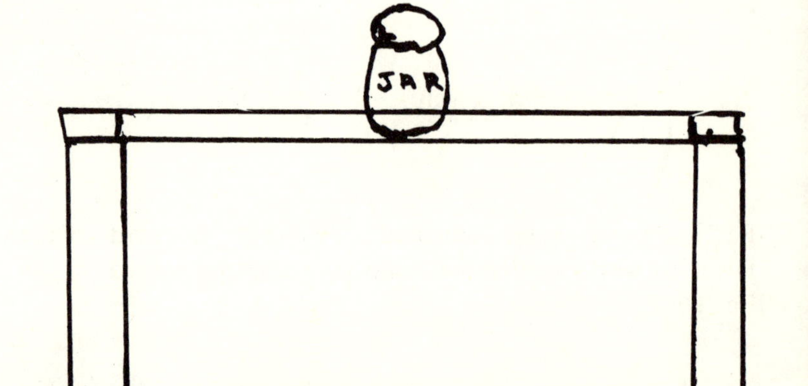

John saved his money and kept it in this jar, until he had enough to buy his bees and equipment.

John now waited for his honeybees to arrive. In the month of May, his neighbor came over and told John his bees were down at the railroad station. John rushed outside to find his father, who would ride John to the station. The excitement went though John like a ball of fire. His father drove down in his 1935 Plymouth car. When they arrived, the man at the railroad station asked John to remove the bees as soon as possible. He seemed afraid that the bees would get out of their cage. John assured him in no way would the honeybees escape.

John arrived at the train station with his father, anxious to pick up his bees.

On his way home, John was thinking how he would put his bees and the Queen down inside of the hive. After he was home, John removed the top of the beehive and down on the bottom is where John put the cage that had the bees in it. He removed the syrup can which is in the center of the bee cage. This is food for the bees which must be fed during their trip up from the southern states. Off of the cage is a small, square little box that holds the Queen and about four honeybees and one or two drones.

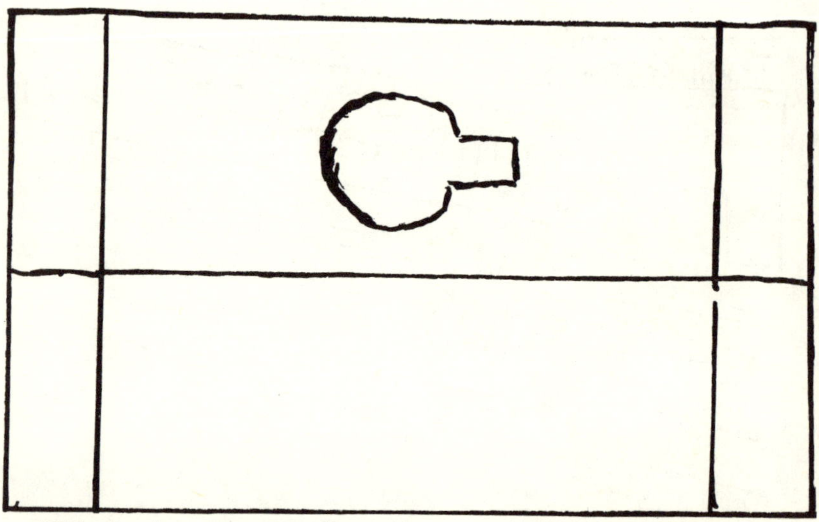

This is the cage that honeybees are shipped in from the southern states. In the center of the cage is a can of syrup for feeding the bees.

John noticed alongside of the syrup can, is a small cage which has a queen inside and about three honeybees and two drones. This little cage is removed from the big cage and placed on top of the frames in the bee hive. On one end of the small cage is a lump of sugar. After the honeybees are released from the big cage, they move up on the frames to the queen. Honeybees will eat the sugar until the queen is free to come out of the small cage. The queen will then move down on the honeycomb.

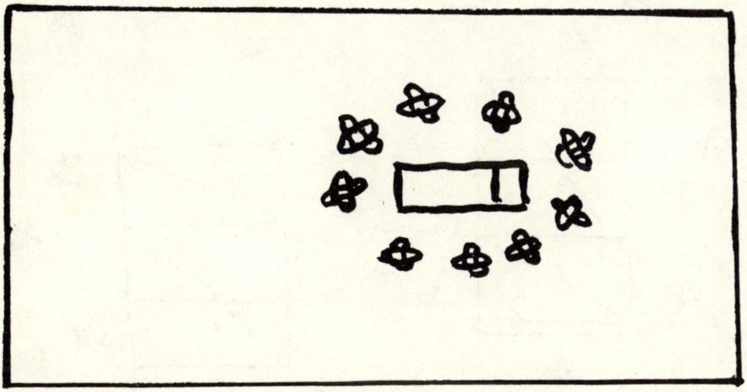

The honeybees are all over the little cage. The honeybees are eating a lump of sugar on one end of the cage to release the queen inside.

Each day while in school, John could not wait to get home. He wanted to watch the honeybees coming back with pollen on their hind legs. The workers would have different colors on their backs and hind legs. These colors came from the different types of flowers that they work. The bees carried on right up to dark. Then all you could see were little guards in the front of their hive entrance.

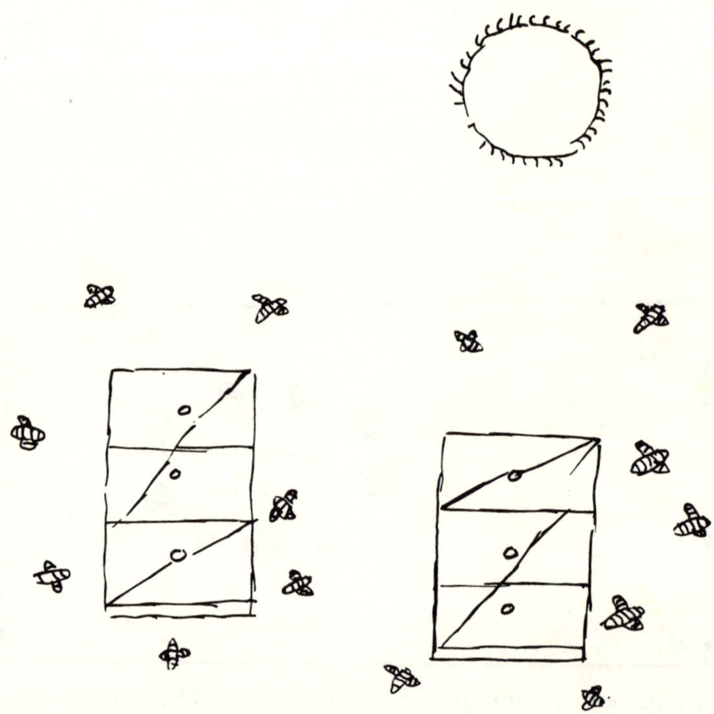

The honeybees are flying back to the hives with pollen and nectar from the flowers.

John noticed toward evening that the bees were fanning their wings in the front entrance. John learned after reading his book that whenever the bees are fanning inside or outside the front entrance two things are taking place. One, the bees are cooling the hive down and two, the bees are evaporating the water from the nectar. There is about 80 percent water carried in the nectar. This is removed by the bees, which in time turns the nectar into honey.

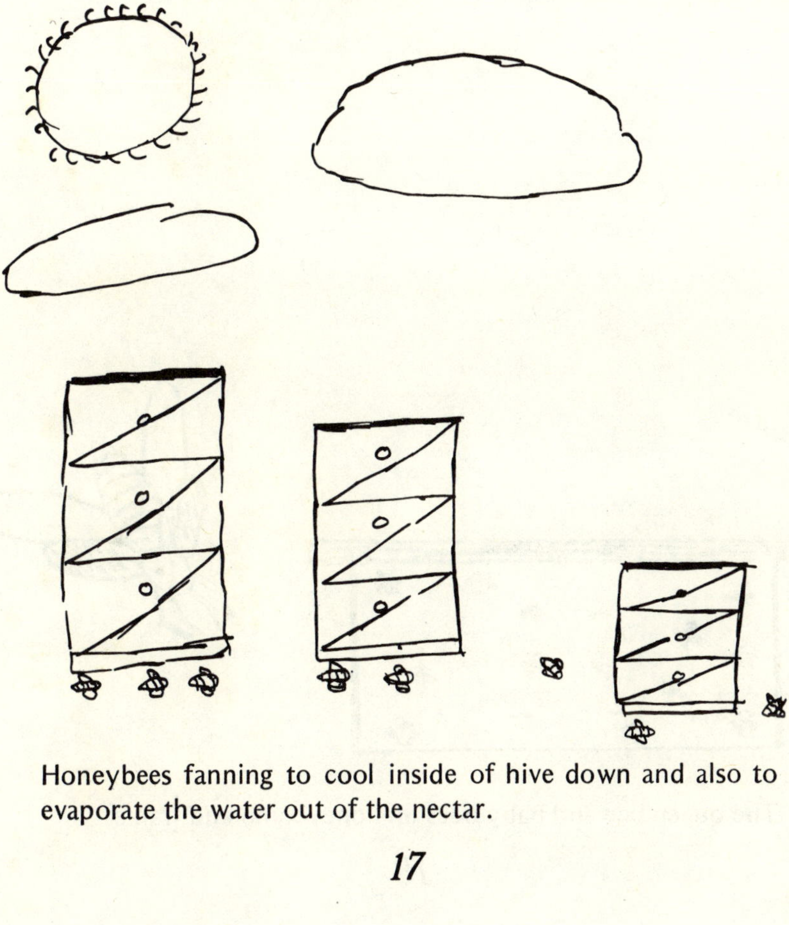

Honeybees fanning to cool inside of hive down and also to evaporate the water out of the nectar.

As the summer went by, John learned more through reading his bee book and working down inside with the bees in the hive. When the queen lays her eggs, in twenty-four to forty-eight hours the little egg drops down to the bottom of the cell where it turns to a small grub. This grub is fed royal jelly for the first three days, then fed pollen and honey. Before the bee is capped over, nurse bees will feed between 10,000 to 12,000 times. There will be enough food inside so that the young bee will be able to live until she eats her way out of the honeycomb.

The queen bee and baby bees on honeycomb and frame.

After reading a book on honey bees, John learned that the honey bee takes twenty-one days to hatch out, the drone bee twenty-four days and the queen fourteen to sixteen days. When the honey bee comes out she takes about one or two days to clean herself and grow stronger. Then she begins to polish the cells out for the queen, who will lay her egg down inside of the cell. This the worker will do until the eighth day of her life. Then she becomes a nurse bee, feeding young bees and storing nectar and doing guard work.

John Watching Honey Bees Working On Honey Comb.

John examined the honey bees. He discovered the honey bee has five eyes, three on top of her head, and one on each side of her face. She also has four wings, six legs, and two stomachs to carry food back to the hive. On her hind legs are little baskets to help her carry pollen back to the hive where she unloads her pollen into the cells for storage. Under her stomach are eight small holes which open up and produce bee wax. She must eat honey in order to produce bee wax. Some honey bees live in hollow trees and others live in man-made bee hives.

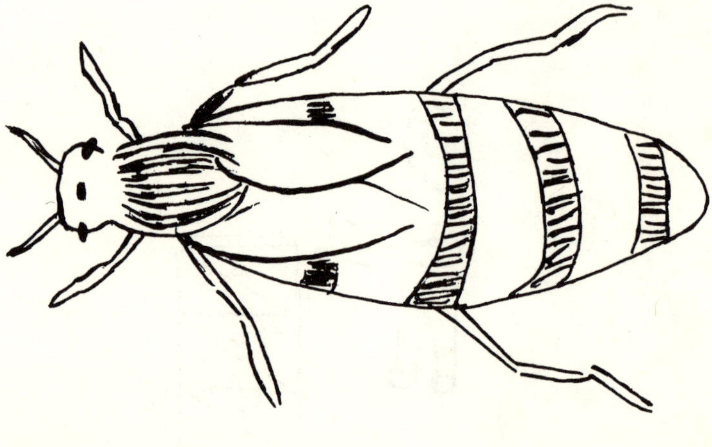

A honeybee with five eyes, four wings, six legs and one stinger.

John watched the honeybees doing a dance on the honeycomb. Sometimes this dance can be seen on the outside of the hive. John read that the honeybee uses the sun as a guide to know which way to go to the flowers. This dance is carried on all day because of the earth moving around the sun.

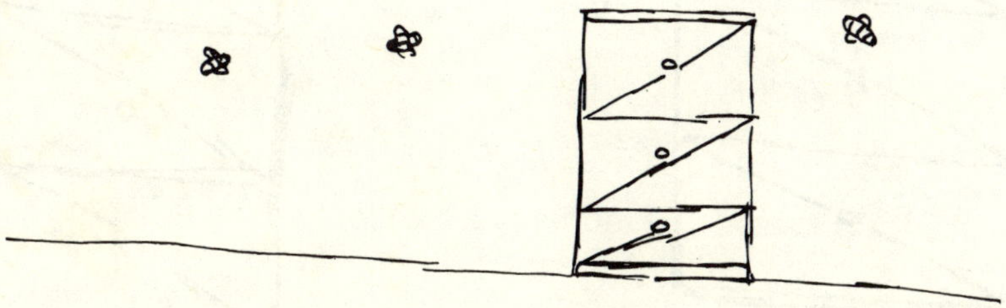

Honeybees must use the sun as a way to guide them to the flowers. The dance they do on the honeycomb is carried out all day.

John made sure each hive had at least sixty-five pounds or more of honey in order to survive the winter. Also, this depends on how many bees are in the hive and how cold the winters are. A hive of bees, at least one year old, can produce fifty-five to 120 pounds of honey for the beekeeper in a season.

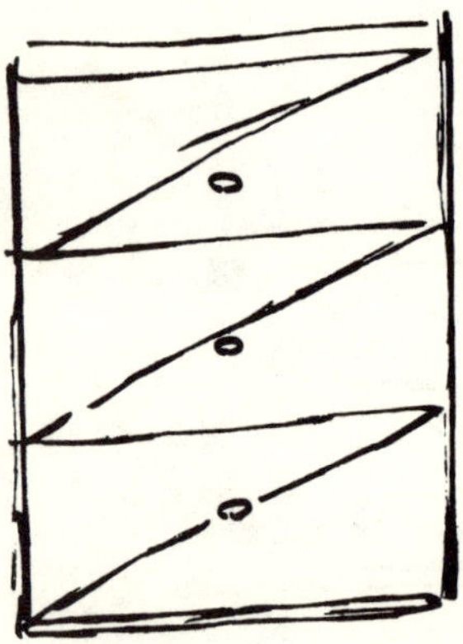

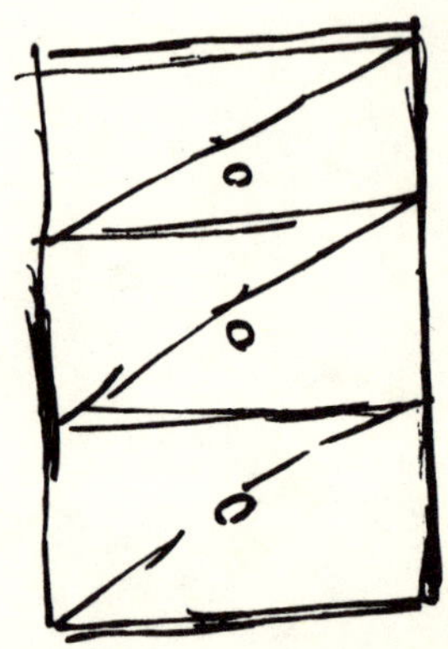

Here are two hives ready for winter. Each one weighs 109 pounds.

As the first season came to an end, John had two beautiful hives. At the end of September, John covers his hives with black felt paper to protect them during the winter.

When Fall came John wrapped his hive in black felt paper to keep them warm during the cold winter months ahead.

As John learned more about bees, people started asking his help removing swarms of honeybees from trees or bushes in their yards. When he had finished the job and had the honeybees in a hive, most times the lady would give John a dollar or glass of soda. John thought this was real nice of the people.

A swarm of honeybees coming down on a tree in a yard.

John was called by a radio station where a man said there was a swarm of bees on the chimney outside the station. John rode his bicycle and tied the beehive on the back of a small cart attached to the bicycle. When John arrived he found that the honey bees were way down inside the chimney. He tried so hard to get the bees but found there was no way. The man at the radio station asked John if he had caught the swarm of bees. John had to tell him no.

That night when John was home he felt very sad because he wanted the bees very badly.

One day, John received a call from the city police to remove a swarm of honey bees on Main Street. The honey bees were attached to the front part of a car.

John was on his way down to Main Street with his car to pick up the swarm of honey bees.

When he arrived he noticed two hundred to three hundred people waiting for him to come. The Police held the people back so John could start to work. John noticed this was not an ordinary swarm like he had taken away before. At four feet high, the swarm had crossed Main Street and come down on a man's new car. The swarm was on the front license plate and hung down to the ground.

John then went to work, down on his knees, because the bees were on the ground.

John worked hard to find the Queen Bee. When John found her, he placed her on a frame of honeycomb and placed it inside the hive. A man raced over and took a picture of John working with the bees, which later was put in the newspapers. In a short time all the bees followed the Queen Bee, and were soon in the hive. John was cheered by the people for the fine job he had done.

Another man went to John and asked him to talk on a tape recorder, so he could play it back on the radio station that night.

John removed honey bees for many well-known people throughout the Eastern part of the United States. From his valuable experiences, John was asked to teach in different schools and soon found himself doing radio and television appearances.

John takes his children over to the hives to show them where some of the bees are from.

And to bring this story to an end, John likes to look back at all the beautiful things that have happened throughout his life. He feels that in some way he has been rewarded for helping people throughout the years.

The End